Materials in Our World

Developed at
The Lawrence Hall of Science,
University of California, Berkeley
Published and distributed by
Delta Education,
a member of the School Specialty Family

1325235
978-1-60902-028-6
Printing 3 – 7/2012
Quad/Graphics, Leominster, MA

Table of Contents

The Story of a Chair

Here is a kind of chair you might
sit on in a park.
Do you know how it was made?

A chair begins as a **tree** in the **forest**.

A lumberjack cuts the trees
in the forest.

A truck driver takes the trees to the sawmill.
The trees are cut into boards at the sawmill.

The boards go to the lumberyard.
The woodworker buys **wood** at
the lumberyard.

The woodworker makes the chair
from wood.
And then it's put in the park for
you to sit on!

Are You a Scientist?

Scientists ask questions.

Scientists look at things to find out about them.
Scientists touch, smell, and listen to things, too.

Scientists **compare** things.
Scientists look at how things are the same and different.

Scientists talk about what they find out.

Are you a scientist?
You are!

The Story of a Box

This box is strong.
But it is made of **paper**.
Do you know how a box is made?

A box starts in the forest as a tree.
First, the trees are cut.
Then, the trees go to a paper mill.

The trees are chopped to
make **sawdust**.
Sawdust is soaked in **water**.
This makes **pulp**.

The pulp is flattened and dried
to make paper.
The paper is put on a roll.
This roll of paper can be used
to make cardboard.

The paper is layered and
glued together.
This makes cardboard.
The cardboard is cut and folded.

Now the box is ready to use!

What Is Fabric Made From?

Fabric can be colorful.
Fabric can be nubby or smooth.
Fabric can be strong.

Where does fabric come from?

People make fabric.
Some fabric is made from
animal hair or fur.
Wool comes from sheep.

Wool is spun into yarn.
The yarn is dyed to give it color.
The yarn is woven to make wool fabric.

Silk comes from silkworm cocoons.
Silk fibers are pulled from the cocoons.
The fibers are made into thread.

The thread is woven to make silk fabric.

Some fabric is made from plants.
Cotton comes from cotton plants.
Cotton is spun into thread.

The thread is woven to make cotton fabric.

Jute comes from jute plants.
Jute plants are cut and soaked.

Jute is dried and spun into thread. The thread is woven to make burlap fabric.

Not all fabric is made from
animals or plants.
Some fabric is made from **oil**.
Oil is pumped from the ground.

Nylon thread is made from oil. The thread is woven to make nylon fabric.

Some fabric is woven.

Some fabric is knitted.
Look at the fabric you are wearing.
How do you think it was made?

How Are Fabrics Used?

Some fabrics are waterproof.

Waterproof fabrics are good for umbrellas, awnings, and tents.

Some fabrics are easy to wash.

They are good for towels, pants, and shirts.

Some fabrics are colorful.

They are good for flags and dressing up on special days.

Some fabrics are strong.
They are good for backpacks
and shoes.

Some fabrics are soft.
They keep you warm.
Soft fabrics are good for snuggling!

What kinds of fabric are you wearing?

How Are Rocks Different?

Think about a rock.

What are its **properties**?

What does the rock look like?

What size is the rock? Is it big or small?
Is it heavy or light? Can you hold it in your hand?

What is its **texture**? Is it smooth or rough?
What is its **shape**? Is it round or flat?

What color is it? Is it more than one color?

Does it have a pattern? Does it have spots or stripes?

Does it have **luster**? Is it shiny or dull?

Does your rock look different when it gets wet?

How do these rocks compare
to your rock?
What does the rock you're
thinking about look like?

How Are Rocks, Soil, and Water Used?

Rocks are used to make things. Walls and walkways are made of rocks.

Buildings and statues are made of rocks, too.

Soil is used to grow plants and trees. Plants and trees are used for **food** and **shelter**.

Soil is used to make bricks.
Bricks are used to make houses.

Water is used to cook and wash.
Water is used to grow plants.
We also use water to drink.

You can use rocks, soil, and water to make paintings, too!

Land, Air, and Water

Do you ever **recycle** paper?
If you do, you are helping to
save a tree.

We use trees for many things. Trees are important to us. Other things are important to us, too.

Land, **air**, and water are important.
We need good land for growing food.

We need good water to drink.

We need good air to breathe.
How can we take care of our
land, air, and water?

I Am Wood

I Am Wood
Nina Steinberg
♩= 160
mf
Chorus
G C D
I am wood, (clap) (clap) I am wood, (clap) (clap) I come from a
tree. I am wood, (clap) (clap) I am wood, (clap) (clap) come and look for me.
Verse
I could be a floor, - I could be a door, - I could be a ta - ble or chair. I could be a big stick,
I could be - rough, - I could be - tough, - I could be ham-mered and cut. I could give you splin - ters,
I could be a tooth - pick. - You could find me a - ny where. Uh - huh, that's right,
keep you warm in win - ters. Re - plant me or you'll use me up. Re -
You could find me a - ny - where. I am
plant me or you'll use me . up.

Glossary

air a mixture of gases that we breathe **(55)**

compare to look at how things are the same and different **(11)**

fabric a flexible material used to make clothing. Fabric and cloth are the same. **(19)**

food something plants and animals need to live **(49)**

forest a natural area where trees grow **(4)**

jute a plant used to make burlap **(26)**

land the solid surface of Earth **(55)**

luster how shiny or dull a rock looks **(45)**

oil a liquid pumped from the ground. Oil is used by people to make nylon and many other things. **(28)**

paper a flexible material made from wood **(13)**

property something you can observe about an object or material. Size, color, shape, texture, and smell are properties. **(41)**

pulp a mixture of sawdust and water used to make paper **(15)**

recycle to use again **(53)**

sawdust the small bits made when a saw cuts wood **(15)**

scientist a person who studies the natural world **(9)**

shape how round or flat an object is **(43)**

shelter a safe place where people and animals live **(49)**

soil a mix of sand, pebbles, and dead plants **(49)**

texture the way something feels **(43)**

tree a plant with a woody stem, roots, and branches with leaves **(4)**

water the most common liquid on Earth. Water is found in oceans, lakes, and streams. **(15)**

wood a strong material in the trunks and branches of trees **(7)**